ADHD and my struggles with it and how living with it is difficult

Attention deficit disorder

Every year the National Information Center for Children and Youth with Disabilities (NICHCY) receives thousands of requests for information about the education and special needs of children and youth with Attention-Deficit Disorder (ADHD, ADD/ADHD). Over the past several years, ADD or A.D.D. (ADHD, ADD/ADHD) has received a tremendous amount of attention from parents, professionals, and policymakers across the country - so much so, in fact, that nearly everyone has now heard about ADD or A.D.D..

. . . .

While helpful to those challenged by this disability, such widespread recognition creates the possibility of improper diagnostic practice and inappropriate treatment. Now, more than ever, parents who suspect their child might have ADHD (A.D.D., ADD/ADHD, ADD-ADHD, A.D.D.) and parents of children diagnosed with the disorder need to evaluate information, products, and practitioners carefully.

This NICHCY Briefing Paper is intended to serve as a guide to help parents and educators know what ADHD (A.D.D., ADD/ADHD, ADD-ADHD, A.D.D.) is, what to look for, and what to do. While acknowledging that adults, too, can have ADHD (ADD/ADHD, ADD-ADHD, A.D.D.), this paper focuses on the disorder as it relates to children and youth.

. . .

What is Attention-Deficit Disorder?

ADD is officially called Attention-Deficit/Hyperactivity Disorder, or AD/HD (American Psychiatric Association, 1994), although most lay people, and even some professionals, still call it ADD or A.D.D. (the names given in 1980) or ADHD . The disorder's name has changed as a result of scientific advances and the findings of careful field trials; researchers now have strong evidence to support the position that AD/HD [A.D.D. or ADHD] [as we will refer to the disorder throughout the remainder of this Briefing Paper] is not one specific disorder with different variations. In keeping with this evidence, AD/HD [A.D.D. OR ADHD] is now divided into three subtypes, according to the main features associated with the disorder: inattentiveness, impulsivity, and hyperactivity. The three subtypes are:

- AD/HD [A.D.D. OR ADHD] Predominantly Combined Type,
- AD/HD [A.D.D. OR ADHD] Predominantly Inattentive Type, and
- AD/HD [A.D.D. OR ADHD] Predominantly Hyperactive-Impulsive Type.

These subtypes take into account that some children with AD/HD [A.D.D. OR ADHD] have little or no trouble sitting still or inhibiting behavior, but may be predominantly inattentive and, as a result, have great difficulty getting or staying focused on a task or activity. Others with AD/HD [A.D.D. OR ADHD] may be able to pay attention to a task but lose focus because they may be predominantly hyperactive-impulsive and, thus, have trouble controlling impulse and activity. The most prevalent subtype is the Combined Type. These children will have significant symptoms of all three characteristics.

. . .

What Causes AD/HD [A.D.D. OR ADHD]?

AD/HD [A.D.D. OR ADHD] is a neurobiologically-based developmental disability estimated to affect between 3-5% of the school age population (Professional Group for Attention and Related Disorders,1991). No one knows exactly what causes AD/HD [A.D.D. OR ADHD]. Scientific evidence suggests that the disorder is genetically transmitted in many cases and results from a chemical imbalance or deficiency in certain neurotransmitters, which are chemicals that help the brain regulate behavior. In addition, a landmark study conducted by the National Institute of Mental Health showed that the rate at which the brain uses glucose, its main energy source, is lower in subjects with AD/HD [A.D.D. OR ADHD] than in subjects without AD/HD [A.D.D. OR ADHD] (Zametkin et al., 1990).

Even though the exact cause of AD/HD [A.D.D. OR ADHD] remains unknown, we do know that AD/HD [A.D.D. OR ADHD] is a neurologically-based medical problem. Parents and teachers do not cause AD/HD [A.D.D. OR ADHD]. Still, there are many things that both can do to help a child manage his or her AD/HD [A.D.D. OR ADHD]-related difficulties. Before we look at what needs to be done, however, let us look at what AD/HD [A.D.D. OR ADHD] is and how it is diagnosed.

. . .

What Are the Signs of AD/HD [A.D.D. OR ADHD]?

Professionals who diagnose AD/HD [A.D.D. OR ADHD] use the diagnostic criteria set forth by the American Psychiatric Association (1994) in the Diagnostic and Statistical Manual of Mental Disorders; the fourth edition of this manual, known as the DSM-IV, was released in May 1994. The criteria in the DSM-IV (discussed below) and the other essential diagnostic features listed in the box labeled "Defining Attention-Deficit/Hyperactivity Disorder" are the signs of AD/HD [A.D.D. OR ADHD].

As can be seen, the primary features associated with the disability are inattention,hyperactivity, and impulsivity. The discussion below describes each of these features and lists their symptoms, as given in the DSM-IV.

. . .

Inattention

A child with AD/HD [A.D.D. OR ADHD] is usually described as having a short attention span and as being distractible. In actuality, distractibility and inattentiveness are not synonymous. Distractibility refers to the short attention span and the ease with which some children can be pulled off-task. Attention, on the other hand, is a process that has different parts. We focus (pick something on which to pay attention), we select (pick something that needs attention at that moment) and we sustain (pay attention for as long as is needed). We also resist (avoid things that remove our attention from where it needs to be), and we shift (move our attention to something else when needed).

When we refer to someone as distractible, we are saying that a part of that person's attention process is disrupted. Children with AD/HD [A.D.D. OR ADHD] can have difficulty with one or all parts of the attention process. Some children may have difficulty concentrating on tasks (particularly on tasks that are routine or boring). Others may have trouble knowing where to start a task. Still others may get lost in the directions along the way. A careful observer can watch and see where the attention process breaks down for a particular child.

Symptoms of inattention, as listed in the DSM-IV, are:[*]

- often fails to give close attention to details or makes careless mistakes in schoolwork, work, or other activities;
- often has difficulty sustaining attention in tasks or play activities;
- often does not seem to listen when spoken to directly;
- often does not follow through on instructions and fails to finish schoolwork, chores, or duties in the workplace (not due to oppositional behavior or failure to understand instructions);
- often has difficulty organizing tasks and activities;
- often avoids, dislikes, or is reluctant to engage in tasks that require sustained mental effort (such as schoolwork or homework);
- often loses things necessary for tasks or activities (e.g., toys, school assignments,pencils, books, or tools);
- is often easily distracted by extraneous stimuli;
- is often forgetful in daily activities.

_____

[*] (American Psychiatric Association, 1994, pp.83-84)

. . .

Hyperactivity

Excessive activity is the most visible sign of AD/HD [A.D.D. OR ADHD]. The hyperactive toddler/preschooler is generally described as "always on the go" or "motor driven." With age, activity levels may diminish. By adolescence and adulthood, the overactivity may appear as restless, fidgety behavior (American Psychiatric Association, 1994).[*]

Symptoms of hyperactivity, as listed in the DSM-IV, are:

- often fidgets with hands or feet or squirms in seat;

- often leaves seat in classroom or in other situations in which remaining seated is expected;
- often runs about or climbs excessively in situations in which it is inappropriate (in adolescents or adults, may be limited to subjective feelings of restlessness);
- often has difficulty playing or engaging in leisure activities quietly;
- is often "on the go" or often act as if "driven by a motor;"
- often talks excessively.

---

* (APA, 1994, p. 84)

Signs and symptoms

Inattention, hyperactivity (restlessness in adults), disruptive behavior, and impulsivity are common in ADHD.[32][33] Academic difficulties are frequent as are problems with relationships.[32] The symptoms can be difficult to define as it is hard to draw a line at where normal levels of inattention, hyperactivity, and impulsivity end and significant levels requiring interventions begin.[34]

To be diagnosed per DSM-5, symptoms must be observed in multiple settings for six months or more and to a degree that is much greater than others of the same age.[35] They must also cause problems in the person's social, academic, or work life.[35]

Based on the presenting symptom ADHD can be divided into three subtypes: predominantly inattentive, predominantly hyperactive-impulsive, and combined type.[34]

An individual with inattention may have some or all of the following symptoms:[36]

- Be easily distracted, miss details, forget things, and frequently switch from one activity to another
- Have difficulty maintaining focus on one task
- Become bored with a task after only a few minutes, unless doing something enjoyable
- Have difficulty focusing attention on organizing and completing a task or learning something new
- Have trouble completing or turning in homework assignments, often losing things (e.g., pencils, toys, assignments) needed to complete tasks or activities
- Not seem to listen when spoken to
- Daydream, become easily confused, and move slowly

- Have difficulty processing information as quickly and accurately as others
- Struggle to follow instructions

An individual with hyperactivity may have some or all of the following symptoms:[36]

- Fidget and squirm in their seats
- Talk nonstop
- Dash around, touching or playing with anything and everything in sight
- Have trouble sitting still during dinner, school, doing homework, and story time
- Be constantly in motion
- Have difficulty doing quiet tasks or activities

These hyperactivity symptoms tend to go away with age and turn into "inner restlessness" in teens and adults with ADHD.[1]

An individual with impulsivity may have some or all of the following symptoms:[36]

- Be very impatient
- Blurt out inappropriate comments, show their emotions without restraint, and act without regard for consequences
- Have difficulty waiting for things they want or waiting their turns in games
- Often interrupt conversations or others' activities

People with ADHD more often have difficulties with social skills, such as social interaction and forming and maintaining friendships. This is true for all subtypes. About half of children and adolescents with ADHD experience social rejection by their peers compared to 10–15% of non-ADHD children and adolescents. People with ADHD have attention deficits which cause difficulty processing verbal and nonverbal language which can negatively affect social interaction. They also may drift off during conversations, and miss social cues.[37]

Difficulties managing anger are more common in children with ADHD[38] as are poor handwriting[39] and delays in speech, language and motor development.[40][41] Although it causes significant impairment, particularly in modern society, many children with ADHD have a good attention span for tasks they find interesting.[10]

Associated disorders

In children ADHD occurs with other disorders about ⅔ of the time.[10] Some commonly associated conditions include:

- Learning disabilities have been found to occur in about 20–30% of children with ADHD. Learning disabilities can include developmental speech and language disorders and academic skills disorders.[42] ADHD, however, is not considered a learning disability, but it very frequently causes academic difficulties.[42]

- Tourette syndrome has been found to occur more commonly in the ADHD population.[43]
- Oppositional defiant disorder (ODD) and conduct disorder (CD), which occur with ADHD in about 50% and 20% of cases respectively.[44] They are characterized by antisocial behaviors such as stubbornness, aggression, frequent temper tantrums, deceitfulness, lying, and stealing.[45] About half of those with hyperactivity and ODD or CD developantisocial personality disorder in adulthood.[46] Brain imaging supports that conduct disorder and ADHD are separate conditions.[47]
- Primary disorder of vigilance, which is characterized by poor attention and concentration, as well as difficulties staying awake. These children tend to fidget, yawn and stretch and appear to be hyperactive in order to remain alert and active.[45]
- Mood disorders (especially bipolar disorder and major depressive disorder). Boys diagnosed with the combined ADHD subtype are more likely to have a mood disorder.[48]Adults with ADHD sometimes also have bipolar disorder, which requires careful assessment to accurately diagnose and treat both conditions.[49]
- Anxiety disorders have been found to occur more commonly in the ADHD population.[48]
- Obsessive-compulsive disorder (OCD) can co-occur with ADHD and shares many of its characteristics.[45]
- Substance use disorders. Adolescents and adults with ADHD are at increased risk of developing a substance use problem.[1] This is most commonly seen with alcohol orcannabis.[1] The reason for this may be an altered reward pathway in the brains of ADHD individuals.[1] This makes the evaluation and treatment of ADHD more difficult, with serious substance misuse problems usually treated first due to their greater risks.[19][50]
- Restless legs syndrome has been found to be more common in those with ADHD and is often due to iron deficiency anaemia.[51][52] However, restless legs can simply be a part of ADHD and requires careful assessment to differentiate between the two disorders.[53]
- Sleep disorders and ADHD commonly co-exist. They can also occur as a side effect of medications used to treat ADHD. In children with ADHD, insomnia is the most common sleep disorder with behavioral therapy the preferred treatment.[54][55] Problems with sleep initiation are common among individuals with ADHD but often they will be deep sleepers and have significant difficulty getting up in the morning.[56] Melatonin is sometimes used in children who have sleep onset insomnia.[57]

There is an association with persistent bed wetting,[58] language delay,[59] and developmental coordination disorder (DCD), with about half of people with DCD having ADHD.[60]The language delay in people with ADHD can include problems with auditory processing disorders such as short-term auditory memory weakness, difficulty following instructions, slow speed of processing written and spoken language, difficulties listening in distracting environments e.g. the classroom, and weakness in reading comprehension.[61]

Cause

The cause of most cases of ADHD is unknown; however, it is believed to involve interactions between genetic and environmental factors.[62][63] Certain cases are related to previous infection of or trauma to the brain.[62]

Genetics

See also: Hunter vs. farmer hypothesis

Twin studies indicate that the disorder is often inherited from one's parents with genetics determining about 75% of cases.[19][64][65] Siblings of children with ADHD are three to four times more likely to develop the disorder than siblings of children without the disorder.[66] Genetic factors are also believed to be involved in determining whether ADHD persists into adulthood.[67]

Typically, a number of genes are involved, many of which directly affect dopamine neurotransmission.[68][69] Those involved with dopamine include DAT, DRD4, DRD5, TAAR1, MAOA, COMT, and DBH.[69][70][71] Other genes associated with ADHD include SERT, HTR1B, SNAP25, GRIN2A, ADRA2A, TPH2, and BDNF.[68][69] A common variant of a gene called LPHN3 is estimated to be responsible for about 9% of cases and when this variant is present, people are particularly responsive to stimulant medication.[72]

As ADHD is common, natural selection likely favored the traits, at least individually, and they may have provided a survival advantage.[73] For example, some women may be more attracted to males who are risk takers, increasing the frequency of genes that predispose to ADHD in the gene pool.[74] As it is more common in children of anxious or stressed mothers, some argue that ADHD is an adaptation that helps children face a stressful or dangerous environment with, for example, increased impulsivity and exploratory behavior.[75]

Hyperactivity might have been beneficial, from an evolutionary perspective, in situations involving risk, competition, or unpredictable behavior (i.e. exploring new areas or finding new food sources). In these situations, ADHD could have been beneficial to society as a whole even while being harmful to the individual.[74] Additionally, in certain environments it may have offered advantages to the individuals themselves, such as quicker response to predators or superior hunting skills.[76]

People with Down syndrome are more likely to have ADHD.[77]

Environment

See also: Diet and attention deficit hyperactivity disorder

Environmental factors are believed to play a lesser role. Alcohol intake during pregnancy can cause fetal alcohol spectrum disorders which can include ADHD or symptoms like it.[78] Exposure to tobacco smoke during pregnancy can cause problems with central nervous system development and can increase the risk of ADHD.[79] Many children exposed to tobacco do not develop ADHD or only have mild symptoms which do not reach the threshold for a diagnosis. A combination of a genetic predisposition with tobacco exposure may explain why some children exposed during pregnancy may develop ADHD and others do not.[80] Children exposed to lead, even low levels, or polychlorinated biphenyls may develop problems which resemble ADHD and fulfill the diagnosis.[81] Exposure to the organophosphate insecticides chlorpyrifos and dialkyl phosphate is associated with an increased risk; however, the evidence is not conclusive.[82]

Very low birth weight, premature birth and early adversity also increase the risk[83] as do infections during pregnancy, at birth, and in early childhood. These infections include, among others, various viruses (measles, varicella, rubella, enterovirus 71) and streptococcal bacterial infection.[84] At least 30% of children with a traumatic brain injury later develop ADHD[85] and about 5% of cases are due to brain damage.[86]

Some children may react negatively to food dyes or preservatives.[87] It is possible that certain food coloring may act as a trigger in those who are genetically predisposed but the evidence is weak.[88]:452 The United Kingdom and European Union have put in place regulatory measures based on these concerns; the FDA has not.[89]

Society

The diagnosis of ADHD can represent family dysfunction or a poor educational system rather than an individual problem.[90] Some cases may be explained by increasing academic expectations, with a diagnosis being a method for parents in some countries to get extra financial and educational support for their child.[86] The youngest children in a class have been found to be more likely to be diagnosed as having ADHD possibly due to their being developmentally behind their older classmates.[91][92] Behavior typical of ADHD occurs more commonly in children who have experienced violence and emotional abuse.[19]

Per social construction theory it is societies that determine the boundary between normal and abnormal behavior. Members of society, including physicians, parents, and teachers, determine which diagnostic criteria are used and, thus, the number of people affected.[93] This leads to the current situation where the DSM-IV arrives at levels of ADHD three to four times higher than those obtained with the ICD-10.[16] Thomas Szasz, a supporter of this theory, has argued that ADHD was "invented and not discovered."[94][95]

Pathophysiology

Current models of ADHD suggest that it is associated with functional impairments in some of the brain's neurotransmitter systems, particularly those involving dopamine and norepinephrine.[96] The dopamine and norepinephrine pathways that originate in the ventral tegmental area and locus coeruleus project to diverse regions of the brain and govern a variety of cognitive processes.[97] The dopamine pathways and norepinephrine pathways which project to the prefrontal cortex and striatum are directly responsible for modulating executive function (*cognitive control* of behavior), motivation, reward perception, and motor function;[96][97] these pathways are known to play a central role in the pathophysiology of ADHD.[97][98][99] Larger models of ADHD with additional pathways have been proposed.[96][98][99]

Brain structure

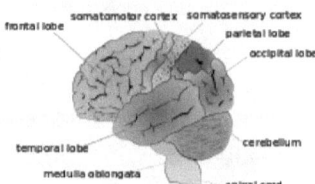

Diagram of the human brain

In children with ADHD, there is a general reduction of volume in certain brain structures, with a proportionally greater decrease in the volume in the left-sided prefrontal cortex.[96][100] The posterior parietal cortex also shows thinning in ADHD individuals compared to controls.[96] Other brain structures in the prefrontal-striatal-cerebellar and prefrontal-striatal-thalamic circuits have also been found to differ between people with and without ADHD.[96][98][99]

Neurotransmitter pathways

Previously it was thought that the elevated number of dopamine transporters in people with ADHD was part of the pathophysiology but it appears that the elevated numbers are due to adaptation to exposure to stimulants.[101] Current models involve the mesocorticolimbic dopamine pathway and the locus coeruleus-noradrenergic system.[96][97] ADHD psychostimulants possess treatment efficacy because they increase neurotransmitter activity in these systems.[96][97][102] There may additionally be abnormalities in serotoninergic and cholinergic pathways.[102][103] Neurotransmission of glutamate, a cotransmitter with dopamine in the mesolimbic pathway,[104] seems to be also involved.[105]

Executive function and motivation

ADHD symptoms involve a difficulty with executive functions.[56][97] Executive function refers to a number of mental processes that are required to regulate, control, and manage daily life tasks.[56][97] Some of these impairments include problems with organization, time keeping, excessive procrastination, concentration, processing speed, regulating emotions, and utilizing working memory.[56] People usually have decent long-term memory.[56] The criteria for an executive function deficit are met in 30–50% of children and adolescents with ADHD.[106] One study found that 80% of individuals with ADHD were impaired in at least one executive function task, compared to 50% for individuals without ADHD.[107] Due to the rates of brain maturation and the increasing demands for executive control as a person gets older, ADHD impairments may not fully manifest themselves until adolescence or even early adulthood.[55]

ADHD has also been associated with motivational deficits in children.[108] Children with ADHD find it difficult to focus on long-term over short-term rewards, and exhibit impulsive behavior for short-term rewards.[108] In these individuals, a large amount of positive reinforcement effectively improves task performance.[108] ADHD stimulants may improve persistence in ADHD children as well.[108]

Diagnosis

ADHD is diagnosed by an assessment of a person's childhood behavioral and mental development, including ruling out the effects of drugs, medications and other medical or psychiatric problems as explanations for the symptoms.[19] It often takes into account feedback from parents and teachers[9] with most diagnoses begun after a teacher raises concerns.[86] It may be viewed as the extreme end of one or more continuous human traits found in all people.[19] Whether someone responds to medications does not confirm or rule out the diagnosis. As imaging studies of the brain do not give consistent results between individuals, they are only used for research purposes and not diagnosis.[109]

In North America, the DSM-IV or DSM-5 criteria are often used for diagnosis, while European countries usually use the ICD-10. With the DSM-IV criteria a diagnosis of ADHD is3–4 times more likely than with the ICD-10 criteria.[16] It is classified as neurodevelopmental psychiatric disorder.[114] Additionally, it is classified as a disruptive behavior disorderalong with oppositional defiant disorder, conduct disorder, and antisocial personality disorder.[110] A diagnosis does not imply a neurological disorder.[19]

Associated conditions that should be screened for include anxiety, depression, oppositional defiant disorder, conduct disorder, and learning and language disorders. Other conditions that should be considered are other neurodevelopmental disorders, tics, and sleep apnea.[111]

Diagnosis of ADHD using quantitative electroencephalography (QEEG) is an ongoing area of investigation, although the value of QEEG in ADHD is currently unclear.[112][113] In the United States, the Food and Drug Administration has approved the use of QEEG to evaluate the morbidity of ADHD.[114]

Diagnostic and Statistical Manual

As with many other psychiatric disorders, formal diagnosis is made by a qualified professional based on a set number of criteria. In the United States, these criteria are defined by the American Psychiatric Association in the DSM. Based on the DSM criteria, there are three sub-types of ADHD:[35]

1. ADHD predominantly inattentive type (ADHD-PI) presents with symptoms including being easily distracted, forgetful, daydreaming, disorganization, poor concentration, and difficulty completing tasks.[8][35]
2. ADHD, predominantly hyperactive-impulsive type presents with excessive fidgetiness and restlessness, hyperactivity, difficulty waiting and remaining seated, immature behavior; destructive behaviors may also be present.[8][35]
3. ADHD, combined type is a combination of the first two subtypes.[8][35]

This subdivision is based on presence of at least six out of nine long-term (lasting at least six months) symptoms of inattention, hyperactivity–impulsivity, or both.[115] To be considered, the symptoms must have appeared by the age of six to twelve and occur in more than one environment (e.g. at home and at school or work).[8] The symptoms must be not appropriate for a child of that age[8][116] and there must be evidence that it is causing social, school or work related problems.[115]

Most children with ADHD have the combined type. Children with the inattention subtype are less likely to act out or have difficulties getting along with other children. They may sit quietly, but without paying attention resulting in the child difficulties being overlooked.[medical citation needed]

International Classification of Diseases

In the ICD-10, the symptoms of "hyperkinetic disorder" are analogous to ADHD in the DSM-5. When a conduct disorder (as defined by ICD-10)[40] is present, the condition is referred to as *hyperkinetic conduct disorder*. Otherwise, the disorder is classified as *disturbance of activity and attention*, *other hyperkinetic disorders* or *hyperkinetic disorders, unspecified*. The latter is sometimes referred to as, *hyperkinetic syndrome*.[40]

Adults

Further information: Adult ADHD

Adults with ADHD are diagnosed under the same criteria, including that their signs must have been present by the age of six to twelve. Questioning parents or guardians as to how the person behaved and developed as a child may form part of the assessment; a family history of ADHD also adds weight to a diagnosis.[1] While the core symptoms of ADHD are similar in children and adults they often present differently in adults than in children, for example excessive physical activity seen in children may present as feelings of restlessness and constant mental activity in adults.[1]

Society

The diagnosis of ADHD can represent family dysfunction or a poor educational system rather than an individual problem.[90] Some cases may be explained by increasing academic expectations, with a diagnosis being a method for parents in some countries to get extra financial and educational support for their child.[86] The youngest children in a class have been found to be more likely to be diagnosed as having ADHD possibly due to their being developmentally behind their older classmates.[91] [92] Behavior typical of ADHD occurs more commonly in children who have experienced violence and emotional abuse

I have several symptoms of ADHD and I did suffer through violence and severe abuse in my childhood

I spent many years wondering why I was losing my keys all the time. I was wondering why I cannot get the best grade I can.It is because of my self-forgetting my pen , or being distracted by something at the exam room or losing my pencil. These problems has caused me to get low grades on exams , tests at school, I lost friendships ,not to mention lots of money over being constantly forgetful about my personal items. I embarrassed myself. I sometimes think that I have a brain injury, not ADHD because I lose things every single day. There is not one day that I do not lose my keys, or my bank card or my glasses, clothes etc. but it is Attention deficit disorder.

Early years: I started losing things, at age 6,7 by being forgetful after being beaten across the head too many times when I was young. I went to school and I never did my best, and if I did, it was brief. I would always lose my homework paper or lose my keys and that was trouble for me. ADHD was very problematic for me in sixth grade where I did well on a social studies course but then I failed the final exam due to inattention problems and my grade went from a B to a C. In the seventh grade, I did very bad on my math and science final exams and it had to do with inattention, and fidgeting, In eighth grade ,after reaching the honor roll all year and hardly getting any B's .The inattention, forgetfulness ,and accidents related to ADHD had caused me to get a F on my science final exam and a C on my math (the truth was, was that I had done the science exam in its entirely but the teacher threw out part of my science final exam accidentally because I wrote the final paper on scrap paper and my math final was so easy yet I could not concentrate and my attention span was not keen on that day ,it had to do with ADHD. I thought I was a dumb and I would never be successful in academia. In high school, the effects of add gotten worse. The results were bad grades and, my mental health determination from abuse, and trauma from crime and in my life and I left high school early. I did get my high school GED in 1989.

AS AN ADULT WITH ADHD

I still lose keys almost every day, sometimes permanently. I lose money, glasses, paper and other items on a regular basis and this coincides with other problems I have.

ADHD IN COLLEGE AND ADULTHOOD

STILL THE SAME, UNFORTUNATELY, I continue to lose things every day because of inattention, forgetfulness and it is almost like I black out. Because of this, I became wary of other people and situations because of these cognitive problems .It has caused me to do bad on the LSAT exam, and on a recent law test. I had gotten lower grades during my bachelors and master's degree programs and it robbed me of a healthy life. Despite it, I learned to remain in control of myself, and I remain a highly functional person as a result. I have develop better strategies at dealing with forgetfulness but the effects of ADHD is very painful , But it can be dealt with through determination .TO

ADHD SUFFERERS ,
Please do not let
obstacles from robbing
you of your dreams and
the adhd is an obstacle
that can be dealt with.

www.ingramcontent.com/pod-product-compliance
Lightning Source LLC
Chambersburg PA
CBHW030012190526
45157CB00015B/2456